Contents

Preface..2

About the Author...3

Chapter 1: Introduction..................................4-7

Chapter 2: Immunity Booster Foods.........8-29

References..30-32

Preface

Welcome to Super Immunity Foods. Read on your Kindle device, Smart Phone, Tablet, PC, or Mac. You can also buy this book.

Immunity is the capability of defense against invading microorganisms. Immunity is the ability to protect against infections. Good nutrition can provide better health and prevent disease. This book will guide you which foods can strength your immunity systems.

About the Author

Anupam Rajak received his B.Sc in Botany from the Raghunathpur College, Sidho-Kanho-Birsha University. He has published several articles in international journal. His email address is anupamrajak1234@gmail.com

Chapter 1

Introduction

Immunity:

Immunity is the capability of defense against invading microorganisms. Immunity is the ability to protect against infections. Immunity is the capability of being able to resist a particular disease. Immunity means being protected from invading microorganisms. Microorganisms and viruses are everywhere in our World.

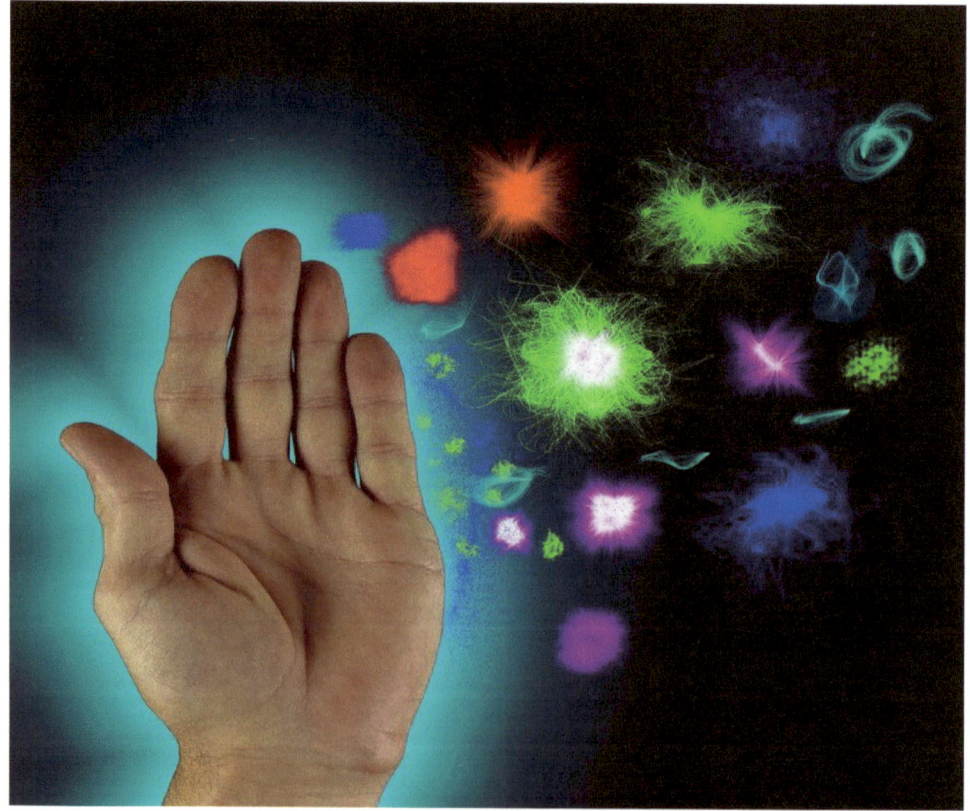

Figure 1. Immune system (Photo Credit: Pixabay)

Types of Immunity:

There are two main types of immunity-

i. Innate Immunity.

ii. Adaptive Immunity.

i. Innate Immunity: Skin is the largest organ in the Human body. Skin prevents the microorganisms and viruses. So, skin is the first line of defense. Many microbes and viruses are enters our body through air, food, or skin. Our body kill the pathogens by phagocytosis mechanisms.

ii. Adaptive Immunity: The function of the adaptive immune response is to destroy invading pathogens.

Nutrition:

We eat food for living. Nutrition may be defined as the utilization of food by living organisms. There are six categories of nutritients:-protein, carbohydrate, fat, fibers, vitamins, minerals, and water. Good nutrition can provide better health and prevent disease.

Figure 2. Nutrition (Photo Credit: Pixabay)

Vitamin:

Vitamins are the micronutrients, which needed in small quantities to human life. Vitamins are fat soluble or water soluble. Different vitamins are different roles.

Fat soluble vitamin:

Fat soluble vitamins includes Vitamin A, Vitamin D, Vitamin E, and Vitamin K.

Water soluble vitamins:

Water soluble vitamins includes Vitamin C and all B Vitamin.

Figure 3. Vitamin C (Photo Credit: Pixabay)

Chapter 2

Immunity Booster Foods

Food is the prime concern for living. Without food, we do not live on Earth. Due to Coronavirus pandemic disease Worldwide, you consume good food to boost your immune system. Some of the following foods to help the immune system-

i. Spinach: Spinach is the leafy green flowering plant, which consists of 91% water, 4% carbohydrates, 3% proteins, Vitamin C, Vitamin A, , Vitamin K, Vitamin E, Vitamin B_6, potassium, manganese, iron, and folate.

Figure 1. Spinach (Photo Credit: Pixabay)

ii. Button mushroom: Button mushroom are nutritious food which contains significant amount of Vitamin B such as riboflavin, niacin, and selenium.

Figure 2. Button mushroom (Photo Credit: Pixabay)

iii. Bell peppers: Bell peppers contains 94% water, 5% carbohydrates, and proteins. They are also contains rich source of Vitamin C, and Vitamin B_6.

Figure 3. Bell Peppers (Photo Credit: Pixabay)

iv. Almonds: Almonds contains rich source of Vitamins B such as riboflavin, and niacin, Vitamin E, and the essential minerals like calcium, copper, iron, magnesium, manganese, phosphorus, and zinc.

Figure 4. Almonds (Photo Credit: Pixabay)

v. Garlic: The scientific name of garlic of Allium sativum. China produces 80% of the garlic.

Figure 5. Garlic (Photo Credit: Pixabay)

vi. Oysters: Oysters contains rich source of Vitamin A and Vitamin B_{12} as well as zinc, iron, calcium, and selenium.

Figure 6. Oysters (Photo Credit: Pixabay)

vii. Watermelon: Watermelon contains 90% of water. Watermelon also contains Vitamin A, Vitamin C, and Vitamin B_1 as well as magnesium, potassium, phosphorus, and calcium.

Figure 7. Watermelon (Photo Credit: Pixabay)

viii. Wheat Germ: Wheat germ is a nutritious food. Wheat Germ contains high amount in Vitamin E and antioxidant properties.

Figure 8. Wheat (Photo Credit: Pixabay)

Ix. Yogurt: Yogurt is a fermented dairy product. We eat Yogurt because it is very nutritious food. Yogurt contains rich source of Vitamin B_{12} and riboflavin, Vitamin D as well as magnesium, and potassium. Yogurt contains high amount of protein and some probiotic bacteria such as Bifidobacteria and Lactobacillus. It may strength our immune system and protect against osteoporosis.

Figure 9. Yogurt (Photo Credit: Pixabay)

x. Tea: Tea is a hot and cold beverages. Tea contains caffeine, theobromin, and theophylline.

Figure 10. Tea (Photo Credit: Pixabay)

xi. Sweet Potato: Sweet potatoes are rich in beta carotene, carbohydrate, and other micronutrients such as Vitamin B_5, and Vitamin B_6.

Figure 11. Sweet Potato (Photo Credit: Pixabay)

xii. Broccoli: Broccoli is an edible vegetable. They are used to treat various disorders like bladder cancer, breast cancer, colorectal cancer, and high cholesterol.

Figure 12. Broccoli (Photo Credit: Pixabay)

xiii. Miso: Miso is a Japanese food produced by fermented soyabeans with salt and koji. Miso contains high amount in protein and rich source of vitamins, and minerals.

Figure 13. Miso (Photo Credit: Pixabay)

xiv. Chicken Soup: Chicken soup has been used to ttreat various disorders like respiratory tract disorders, asthma, and facial pain. Chicken contains high amount of proteins, and also contains vitamins, and minerals, which boost our immunity.

Figure 14. Chicken Soup (Photo Credit: Pixabay)

xv. Pomegranate Juice: Pomegranate juice contains higher level of antioxidant. They are used to treat various disorders like cancer, arthritis, alzheimer's disease, and diabetes. Pomegranate contains good source of vitamins such as Vitamin A, Vitamin C, and Vitamin E.

Figure 15. Pomegranate juice (Photo Credit: Pixabay)

xvi. Elderberry: Elderberry is belongs to the family Adoxaceae. Elderberry contains rich source of Vitamins , which boost our immune systems. They also contains rich source of dietary fibers, phenolic acids, flavonol, and anthocyanins.

Figure 16. Elderberry (Photo Credit: Pixabay)

xvii. Acai Berry: Acai berry contains Vitamin A, calcium, fiber, and sugar.

Figure 17. Acai Strawberry Ice (Photo Credit: Pixabay)

xviii. Citrus: Citrus is belongs to the family Rutaceae. They are rich source of Vitamin C, and fiber. They may protect our brain or fight against cancer.

Figure 18. Citrus Fruit (Photo Credit: Pixabay)

xix. Turmeric: Turmeric has antioxidant and antiinflammatory properties.

Figure 19. Turmeric (Photo Credit: Pixabay)

xx. Kiwifruit: Kiwifruit contains high amounts of Vitamin C and Vitamin K.

Figure 20. Kiwi Fruit (Photo Credit: Pixabay)

xxi. Walnuts: Walnuts are nutritious food. They contains fats, fibers, vitamins, and minerals. They contains rich source in antioxidant, omega-3-fat. They are used to treat various disorders like cancer, alzeihmer's disease, and type-2-diabetes.

Figure 21. Walnuts (Photo Credit: Pixabay)

xxii. Kefir: Kefir is a fermented milk drink. Kefir contains vitamins, dietary minerals, essential amino acids, calcium, iron, phosphorus, magnesium, potassium, sodium, copper, manganese, and zinc.

Figure 22. Kefir (Photo Credit: Pixabay)

References:

1. Alberts B, Johnson A, Lewis J, et al. Molecular Biology of the Cell. 4th edition. New York: Garland Science; 2002. Chapter 24, The Adaptive Immune System. Available from: https://www.ncbi.nlm.nih.gov/books/NBK21070/

2. https://www.timesnownews.com/health/article/coronavirus-prevention-add-these-5-foods-to-your-daily-diet-to-boost-your-immune-system-reduce-your-risk/629309

3. https://en.wikipedia.org/wiki/Bell_pepper

4. https://en.wikipedia.org/wiki/Almond

5. Immune-Boosting Foods: Berries, Oysters, & More. (n.d.). Retrieved August 09, 2020, from https://www.webmd.com/cold-and-flu/ss/slideshow-immune-foods

6. Schend, J. (2020, April 30). What to Eat and Drink to Boost Your Immune System. Retrieved August 09, 2020, from https://www.healthline.com/health/food-nutrition/foods-that-boost-the-immune-system

7. https://www.health.com/food/immunity-boosting-foods

8. https://en.wikipedia.org/wiki/Oyster

9. https://www.medicalnewstoday.com/articles/266886#nutrition

10. https://www.healthline.com/health/wheat-germ-benefits#Are-there-any-side-effects?

11. https://www.organicfacts.net/health-benefits/cereal/health-benefits-of-wheat-germ.html

12. https://www.healthline.com/nutrition/7-benefits-of-yogurt#section6

13. https://en.wikipedia.org/wiki/Tea